ABOVE: *Ryhope pumping station, Sunderland, now an engine museum, houses two massive 100-hp beam engines by Hawthorn of Newcastle, installed in 1869. They provided water for Sunderland until 1967 and originally cost £9,000.*

FRONT COVER: *A six-column engine of 1864 built by Easton, Amos & Co and of a type much used for powering small workshops. It is a compound engine with cylinders of 10 inches and 17 inches bore and the beam is just under 10 feet long. The fluted columns are carefully designed in the correct Greek architectural tradition and the whole engine is of superlative quality. It has been re-erected at the City of Birmingham Museum of Science and Industry, where it is steamed two days each month.*

BEAM ENGINES

T. E. Crowley

Shire Publications Ltd

CONTENTS

Printed by C. I. Thomas & Sons (Haverfordwest) Ltd, Press Buildings, Merlins Bridge, Haverfordwest, Dyfed, Wales.

The valve operating gear of the great 100-inch Cornish engine at Kew Bridge pumping station, London. This beam engine, built by Harvey's, is the largest in Britain; it worked from 1871 to 1944 and moved ten million gallons every twenty-four hours. It is preserved and may be seen any weekend.

The London sewage-pumping station at Crossness has the greatest concentration of beam-engine power on one site. Four James Watt engines of 1864 were converted to triple-expansion working by Goodfellow in 1899. The photograph shows the loft with the four great beams.

INTRODUCTION

It is difficult to realise that the steam engine was invented during the reign of Queen Anne (1702-14). Although the first 'fire engines', as the awed public called them then, probably had hand-operated valves, it was not long before the mechanism was rendered fully automatic.

Very little was known in those days about the behaviour of steam or, indeed, of the properties of metals; no power had been available to man except that of muscle, wind and water, and to design and produce a usable, reliable engine under such circumstances was the work of one of the most original minds the world has known.

The chief need for such an engine was to drive pumps to keep mines free from water, in order that the rich sources of coal and ore could be tapped at lower levels. Steam engines were thus able to release the supplies of raw material and in due course to drive the mills, feed the canals, operate the docks, supply the water and provide the transport, all of which were needed before Britain could step forward into the new world of the Industrial Revolution.

3

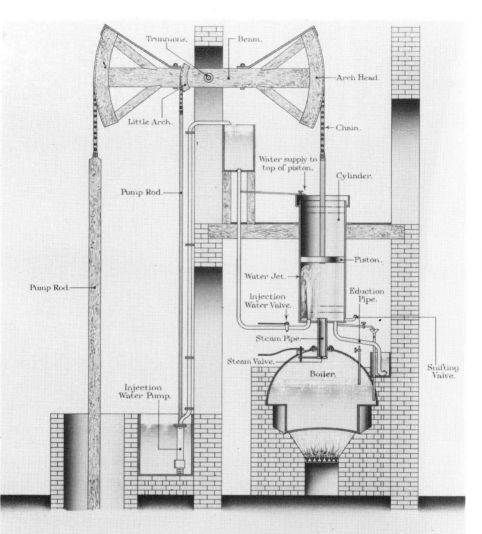

Trunnions.   Beam.

Arch Head.

Little Arch.

Chain.

Water supply to top of piston.

Cylinder.

Pump Rod.

Piston.

Water Jet.

Injection Water Valve.

Eduction Pipe.

Pump Rod.

Steam Pipe.

Steam Valve.

Shifting Valve.

Boiler.

Injection Water Pump.

This diagram shows a typical atmospheric pumping engine, as made by Newcomen, in section, with the piston in the middle of the downward or working stroke. Steam is generated at atmospheric pressure in the boiler and fills the cylinder during the upward stroke of the piston. The steam valve is then closed and the steam is condensed by a jet of cold water causing a vacuum under the piston. The atmospheric pressure acting on the top of the piston forces it down, hence the name "atmospheric" engine, and this constitutes the working stroke. The piston is raised again by the overbalancing weight of the pump-rods.

In 1726 Newcomen engines were already in use for pumping London's water supplies from the Thames. Further engines for the same purpose were installed at Pimlico in 1742 and 1747, as shown here. The site is now occupied by the Grosvenor Hotel.

THOMAS NEWCOMEN

Not much is known about the modest Devonian whose genius was responsible for making possible the Industrial Revolution. Thomas Newcomen (1663-1729) was an ironmonger and brassfounder of Dartmouth, a man of so little importance in the social life of his time that his contemporaries found it hard to believe that he was responsible for the engines of unprecedented power which so soon became essential to any mine of importance both in Britain and other countries. Legal tangles robbed Newcomen of most of the credit and profit which should have been his and he died unrecognised in 1729, by which times scores of his 'fire engines' were at work not only all over Great Britain but in Sweden, Hungary, France, Germany and Belgium. Many of them gave well over a hundred years of service.

The basis of Newcomen's design was this. He was unable to use steam under pressure, since no suitable boiler existed, but he was aware of the pressure of the atmosphere, which is about 15 pounds per square inch. If an upright cylinder, with a sliding piston, was filled with steam which was then condensed, a vacuum would be formed and the pressure of the atmosphere would drive the piston down to the bottom of the cylinder with great force.

Newcomen experimented for about fourteen years with many designs of engine which he made himself with the aid of his devoted assistant, Calley. His main trouble was in getting the steam to condense rapidly enough, but one day a tiny flaw in his brass cylinder allowed a spray of water to leak through to the inside and the piston came down with unprecedented force and speed. This provided the necessary clue, and a water-jet was arranged within the cylinder, to operate as soon as the piston was at the top end of the stroke.

Newcomen arranged a great wooden beam, pivoted in the middle, on the specially strengthened wall of the engine-house, and to one end of the beam he attached the piston by a length of chain. The cylinder was mounted vertically, mouth upwards; to the other end of the beam, again hanging by a chain, Newcomen attached the heavy rods fixed down the mineshaft to the pump at the lowest level. The weight of the rods pulled up the piston in the cylinder and sucked in the steam from the boiler mounted below; then the steam was condensed by the water-jet, the piston descended to fill the vacuum and the pumprods were raised, pumping up the water. The valve admitting the steam to the cylinder and that controlling the water-jet were worked from a rod hung from the beam. Water covered the top of the piston to help seal the gap between piston and cylinder since there was no machine tool available to produce a smooth cylinder bore that was sufficiently accurate.

The first engine erected by Newcomen of which record survives pumped a mine at Dudley Castle, West Midlands, from 1712. As can be seen in the 1719 print, the cylinder was mounted over the boiler and the valve gear is already self-acting.

REFERENCES

By Figures, to the several Members.

1	The Fire Mouth under the Boyler with a Lid or Door.
2	The Boyler 5 Feet, 6 Inches Diameter, 6 Feet 1 Inch high, the Cylindrical part 4 Feet 4 Inches, Content near 13 Hogsheads.
3	The Neck or Throat betwixt the Boyler and the Great Cylinder.
4	A Brass Cylinder 7 Feet 10 Inches high, 21 Inches Diameter, to Rarifie and Condense the Steam.
5	The Pipe which contains the Buoy, 4 Inches Diameter.
6	The Master Pipe that Supplies all the Offices, 4 Inches Diameter.
7	The Injecting Pipe fill'd by the Master Pipe 6, and stopp'd by a Valve.
8	The Sinking Pipe, 4 Inches Diameter, that carries off the hot Water or Steam.
9	A Replenishing Pipe to the Boyler as it wastes with a Cock.
10	A Large Pipe with a Valve to carry the Steam out of Door.
11	The Regulator moved by the 2 Y y and they by the Beam, 12.
12	The Sliding Beam mov'd by the little Arch of the great Beam.
13	Scoggen and his Mate who work Double to the Boy, Y is the Axis of him.
14	The great Y that moves the little y and Regulator, 15 and 11 by the Beam 12
15	The little y, guided by a Rod of Iron from the Regulator.
16	The Injecting Hammer or F that moves upon it's Axis in the Barge 17.
17	Which Barge has a leaking Pipe, besides the Valve nam'd in Nº 7.
18	The Leaking Pipe 1 Inch Diameter, the Water falls into the Well.
19	A Snifting Bason with a Cock, to fill or cover the Air Valve with Water.
20	The Waste Pipe that carries off the Water from the Piston.
21	A Pipe which covers the Piston with a Cock.
22	The Great Sommers that Support the House and Engine.
23	A Lead Cystern, 2 Feet square, fill'd by the Master Pipe 6.
24	The Waste Pipe to that Cystern.
25	The Great Ballanc'd Beam that Works the whole Engine.
26	The Two Arches of the Great Balanced Beam.
27	Two Wooden Frames to stop the Force of the Great Ballanced Beam.
28	The Little Arch of the Great Ballanc'd Beam that moves the Nº 12.
29	Two Chains fix'd to the Little Arch, one draws down, the other up.
30	Stays to the great Arches of the Ballanc'd Beam.
31	Strong Butts of Iron which go through the Arches and secure the Chains.
32	Large Pins of Iron going through the Arch to stop the Force of the Beam.
33	Very strong Chains fixed to Piston and the Plugg and both Arches.
34	Great Springs to stop the Force of the Great Ballanc'd Beam.
35	The Stair-Case from Bottom to the Top.
36	The Ash-hole under the Fire, even with the Surface of the Well.
37	The Door-Case to the Well that receives the Water from the Level.
38	A Stair-Case from the Fire to the Engine and to the Great Door-Case.
39	The Gable-End the Great Ballanc'd Beam goes through.
40	The Colepit-mouth 12 Feet or more above the Level.
41	The dividing of the Pump work into halves in the Pit.
42	The Mouth of the Pumps to the Level of the Well.
43	The Pump-work within the Pit.
44	A Large Cystern of Wood 25 Yards or half way down the Pit.
45	The Pump within the House that Furnishes all the Offices with Water.
46	The Floor over the Well.
47	The Great Door-Case 6 Feet square, to bring in the Boyler.
48	Stays to the Great Frame over the Pit.
49	The Wind to put them down gently or safely.
50	A Turn-Barrel over the Pit, which the Line goes round, not to slip.
51	The Gage-Pipe to know the Depth of the Water within the Boyler.
52	Two Cocks within the Pit to keep the Pump work moist.
53	A little Bench with a Bass to rest when they are weary.
54	A Man going to Replenish the Fire.
55	The Peck-Ax and Proaker.
56	The Centre or Axis of the Great Ballanc'd Beam.

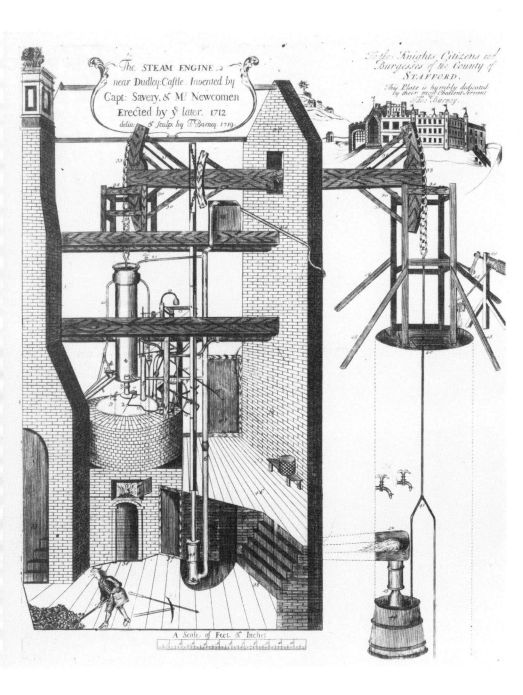

The STEAM ENGINE near Dudley Castle Invented by Capt: Savery, & Mr Newcomen Erected by ye later. 1712 delin: & sculp: by T. Barney. 1719.

To the Knights, Citizens and Burgesses of the County of STAFFORD. This Plate is humbly dedicated by their most Obedient Servant Tho: Barney.

A Scale of Feet & Inches

DEVELOPMENTS

The first Newcomen engine of which details remain to us was erected at Dudley Castle, West Midlands, on a site now lost, in 1712. It had a cylinder 21 inches in diameter and was capable of bringing 10 gallons of water to the adit from a depth of 153 feet at each stroke. It was an immediate success and was rapidly followed by many others, principally for coal-mines, where there was plenty of unsalable fuel to fire the very wasteful boilers. Such engines cost little to run in spite of their inefficiency. In Cornwall, however, where coal had to be transported long distances and landing taxes paid on it, running costs of the engines were found to be almost prohibitive and few were used until design improved.

Abraham Darby of Coalbrookdale was quick to see the potential of the new invention and in quite a short time his foundries were producing pipework, pumps and fittings of all sorts. A few years later, when larger cylinders were needed, he was able to supply them in iron, bored with improved accuracy, and the more expensive brass was used no more. As early as 1752 an engine of 47-inch diameter was at work at Throckley Colliery and, ten years later, one of no less than 74 inches at Walker Colliery, fed by four boilers.

The Hornblower brothers were engine-builders in the middle of the eighteenth century; Jonathan Hornblower built pumping installations to a very high standard of workmanship in Cornwall, where economy was the first consideration, and his brother Josiah crossed the Atlantic with a shipload of parts to introduce the Newcomen engine to America. His first was built at Arlington Copperworks, Maine, in 1753-5, and its cylinder survives in the Smithsonian collection. Josiah had such a bad crossing of the Atlantic that he swore he would never risk another; nor did he, and he died at Belleville in 1809.

Familiarly known as 'Fairbottom Bobs', this ancient Newcomen engine was erected about 1760 and used until 1827 to pump water from a colliery near Ashton-under-Lyne. It survived neglected in its original condition for over a hundred years until Henry Ford bought it for his museum at Dearborn, USA.

'Old Bess', one of the earliest engines built by James Watt, was installed experimentally at the Soho Works in 1777 to pump water. The piston, seen withdrawn, is 33 inches in diameter. The engine is now in the Science Museum.

JOHN SMEATON

A truly versatile engineer — he built the third Eddystone lighthouse — John Smeaton (1724-94) was a worthy successor to Newcomen and did much to improve the design of contemporary engines. He was a practical scientist who despised the trial-and-error methods of the day and concentrated on obtaining more power from smaller engines rather than continually increasing the size of everything. He devised methods of measuring an engine's efficiency and evolved what he called the 'duty', which was the amount of water in millions of pounds which could be raised one foot high by a bushel of coal (94 pounds) consumed. This gave a standard measure by which engines could be compared and it can be translated directly to pounds of fuel per horsepower hour.

Smeaton was associated with the founding of the Carron Ironworks and was able greatly to improve the design of engine parts available from 1760. He increased the size of steampipes, finding that this alone gave improved results, and was able to develop the controls of the engine to vary its speed at will, and according to the load. He developed the 'cataract', a weighted lever with a cup at the end enclosed in a water tank. Water flowed from a tap into the cup and, when full, the weight of it tilted the lever, opening the steam valve and emptying the cup. Thus, by regulating the tap, the speed of the engine could be varied.

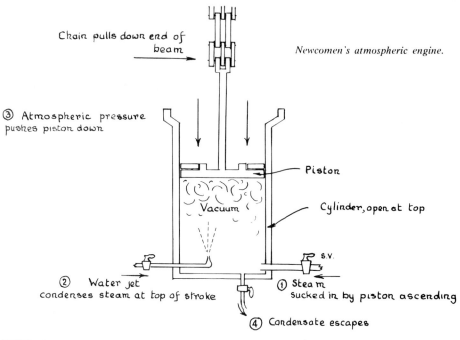

Chain pulls down end of beam

Newcomen's atmospheric engine.

③ Atmospheric pressure pushes piston down

Piston

Vacuum

Cylinder, open at top

S.V.

② Water jet condenses steam at top of stroke

① Steam sucked in by piston ascending

④ Condensate escapes

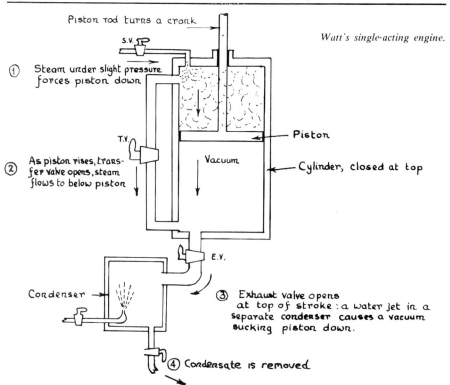

Piston rod turns a crank

S.V.

Watt's single-acting engine.

① Steam under slight pressure forces piston down

T.V.

Piston

② As piston rises, transfer valve opens, steam flows to below piston

Vacuum

Cylinder, closed at top

E.V.

Condenser

③ Exhaust valve opens at top of stroke : a water jet in a separate condenser causes a vacuum sucking piston down.

④ Condensate is removed

Smeaton experimented with laminated beams made up of layers of wood in an attempt to combat the warping which frequently reduced engine efficiency, but he found the design not worth developing: he had more success with improvements to the shape of the cylinder bottom and the design of valve gear. His engines were widely used in many places from London, where they pumped up domestic water from the Thames, to Kronstadt in Russia, where they emptied dry docks.

THE SEPARATE CONDENSER

Engines on the Newcomen or 'atmospheric' principle had now been made and used for over sixty years and they invariably worked by pulling on the beam with a chain. (There was then no such thing as a rotating engine, since it was universally thought that to limit the stroke of a piston by means of a connecting rod and crank would have some disastrous result.)

The fertile mind of James Watt (1736-1819) was working on steam engines at this time. By 1765 he had invented the separate condenser, for which he was granted a patent in 1769. Watt had realised that efficiency would be increased substantially if the cylinder could be kept hot instead of be-

It must be unique for a beam engine to be dug up. The big Newcomen engine was installed at the Reel Fitz Colliery, Workington, in 1779, and after bankruptcy in 1781 most of it was apparently buried. Here is the 5-foot diameter piston, now to be seen in the Greater Manchester Museum of Science and Industry.

ing cooled at every stroke by the water spray needed to condense the steam; he proceeded to design a separate condenser below the cylinder, a cast-iron box which could be kept cold in a tank of water. An extra valve was arranged at the bottom end of the cylinder so that the exhausted steam could enter the condenser, where a water-jet would reduce it to warm water and the resulting vacuum caused the engine to operate. A little pump would keep the condenser from filling up, removing both water and air (there is always some surplus air in steam), and if required the water could be used to replenish the boiler, which did not have to heat it up from cold — another economy. Cylinders were kept hot by means of a steam jacket inside an insulated case.

In 1774 Watt entered into partnership with Matthew Boulton, an established manufacturer of metal goods at the Soho works, Birmingham, and was thus provided with ideal conditions for carrying on his experiments. Boulton and Watt decided to enter the steam engine business in 1775, first as consultants and gradually becoming manufacturers. The factory was powered by means of a great waterwheel, and one of the first engines to be built, 'Old Bess', still on an experimental basis, was principally used in times of drought to pump water from the tailrace back to the headrace and so to keep the wheel going. The step from this to powering the works with steam engines was a big one. Old Bess worked until 1848, when the works closed; most of it is now in the Science Museum.

ROTATIVE ENGINES

For seventy years the steam engine had been thought of primarily as a means of pumping water, but Watt realised that if he could design a rotative engine this could be put to an enormous variety of uses hitherto unthought of. Waterwheels, with their limited power and their unreliability in times of drought, might be superseded for driving mills and factories of all kinds, blowing furnaces, winding up coal and ores from mines, drawing railway trucks up steep slopes, and so on. The Industrial Revolution was under way, factories were opening up every month, all seeking some better source of power. Boulton clearly saw that this could be provided by rotative steam engines and easily persuaded Watt to experiment along these lines.

In 1779, however, James Pickard, a Birmingham button manufacturer, made his engine turn a crank, which he patented, hanging heavy weights on the beam in order to keep the piston chain tight. The engine worked irregularly and badly, but work it did, and another chapter in the history of steam was opened.

The crank having already been patented, Watt, who was always more interested in receiving than in paying out royalties, decided not to use it. He contrived the 'sun and planet' gear, consisting of two cogwheels held in mesh by a link. One was attached to the end of a connecting rod hung from the outer end of the beam and the other was keyed to the flywheel; by this arrangement, known as 'epicyclic' gearing, the flywheel rotated at twice the speed of the engine and could therefore be made very light.

Boiler design was improving and Watt was able to rely on the use of steam at up to about 4 pounds per square inch. He regarded more than this as quite unnecessary but realised that if he closed the cylinder top he could admit steam alternately to either side of the piston, thus doubling his power strokes. Such a concept was impossible with the old chain arrangement, since the piston now had to exert a positive push and pull on the end of the beam; this could only be done by means of a rod working through a gland in the closed cylinder cover and some contrivance to change the linear movement of this rod to the circular movement of the beam end.

Watt himself knew that the thrust of the piston on the beam had a great tendency to bend the unsupported piston rod and he contrived what he called his 'parallel motion' to guide the rod inflexibly in the path it was required to travel. The top of

The 'Lap' engine is of historical importance, being probably the oldest in existence to embody all of James Watt's inventions. The engine was employed at the Soho Works for seventy years driving 43 metal polishing machines; its rating was nearly 14 hp.

the rod was attached to the beam by a swinging link and two pairs of hinged rods were provided, one pair attached each side of the top of the piston rod and the other pair attached to a part of the framing or to the building itself. The movement of the guide rods is susceptible of a rather beautiful geometrical solution, and it was the invention of which James Watt said he was proudest of all. It can be seen in the Science Museum on the 'Lap' engine, which dates from 1788. This engine is an early example of a rotative engine incorporating all Watt's improvements, including the automatic pendulum governor, consisting of two heavy weights which fly outwards under centrifugal force and close the steam valve if the engine speed rises unduly. This was an adaptation of the windmill governor.

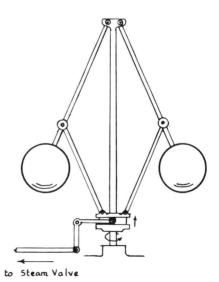

to Steam Valve

Watt's governor.

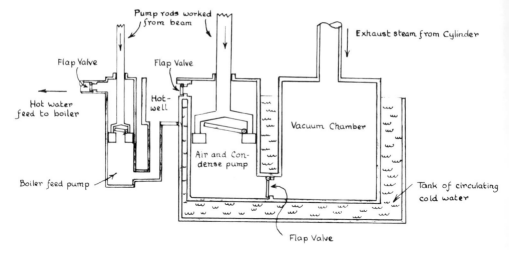

The surface condenser.

PROGRESS IN THE 1790s

There was an immediate demand for the new rotative engines and Watt had a strong control of the market while his patents lasted. By 1800, when they finally expired, the firm and its licensees had put to work about four hundred and fifty engines of all types. Atmospheric engines continued to be built for some years for pumping purposes by people who considered them more reliable and less complex than the Watt engine, especially as the craft of the mechanic scarcely existed. There were, however, one or two people who challenged the monopoly, such as the engineer Adam Heslop, who in 1795 patented a two-cylinder beam engine for which he claimed great advantages. This had a 'hot' cylinder coupled to the wooden beam near the connecting-rod pivot and a 'cold' cylinder immersed in a water tank at the other end. Steam at 1-3 pounds per square inch, after having acted on the piston in the hot cylinder, was led by a valve to the other cylinder and there condensed. The beam

was therefore pushed by Watt's method and pulled by Newcomen's. Watt obtained an injunction although Heslop is known to have erected at least fourteen engines in the Whitehaven area, of which one survives in the Science Museum.

Watt patented the idea of expansive working, whereby the steam admission is cut off early in the stroke, steam continuing to exert pressure on the piston by its own expansion as the pressure drops. This principle is now general in steam engines, but no use was made of it by its patentee, steam pressures at that time being too low. The only great design improvement in the 1790s was made by William Murdock, a foreman at the Boulton and Watt factory, who invented the slide valve to supersede the old sector and drop valves with conical seats. The slide valve was shaped like a D and provided with holes which registered with holes in the cylinder at the correct moments to allow admission and exhaust of the steam.

ABOVE: *This engine has been extensively modified but illustrates a type from the late 1790s, when double-acting cylinders had become common. Note the 'sun and planet' motion in place of the normal crank, and the condenser tank below the cylinder.*

BELOW: *Compare this with the engine above; it shows the progress made in twenty years. Timber is not now used and the beam is of cast iron. There is a normal crank.*

ABOVE LEFT: *The mine pumping engine (1795) at Elsecar, South Yorkshire, is the only Newcomen-type engine remaining in its original position, though now much altered.*

ABOVE RIGHT: *Typical of the valve gear of a Cornish engine, this is fitted to the 1846 Harvey engine at Crofton.*

INTO THE NINETEENTH CENTURY

In 1800 the Boulton and Watt patents finally ran out, although the firm was well constituted to continue engine-building and on the death of its principals was reconstructed and managed by their sons. The firm continued to turn out fine engines until bought up at almost the end of the nineteenth century. After 1800, however, conditions were right for a considerable number of engineers to start designing and constructing a wide variety of designs, many of them highly ingenious. The efficiency of the results increased by leaps and bounds; almost at once timber dis-

appeared from the engine framing, the beam and other moving parts, and for the next half-century or more most engines were constructed almost entirely of cast iron. For the first time, engines appeared which acted directly on a crankshaft, without the interposition of a beam; the cylinder remained vertical in such designs, with the crankshaft above it, since it was generally thought that a horizontal cylinder would necessarily be subject to rapid and uneven wear; the latter type made little headway until the 1850s. Some engines, for example table engines, had the crankshaft

ABOVE: *This tiny engine is located on the Camp Mill Site at Soudley in the Forest of Dean. It has a 10-inch piston and 3-foot stroke and was made in 1805 by Hewlett at the Soudley Foundry to wind coal in the Forest of Dean.*

RIGHT: *For marine use, where headroom was critical, the beam was duplicated and mounted low down each side of the frame, the piston rod actuating them by a pair of return rods. This example at Dumbarton was made by Napier's and drove the Clyde paddle-ship 'Leven' 1824-56. The engine now stands in Dumbarton's new town centre.*

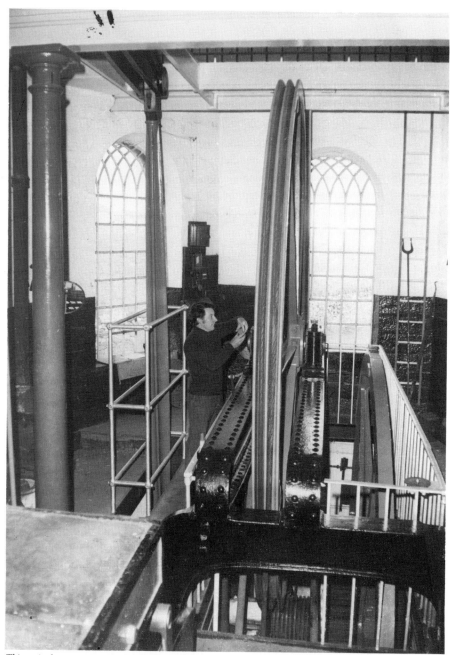

This twin beam engine, by Butterley of Ripley, was built in 1829 to haul wagons up the Middleton Incline (1 in $8\frac{1}{2}$) and worked until 1963.

below the cylinder.

Side-lever designs were produced. This is the name given to the layout in which the beam is situated not above the cylinder but low down each side of the engine framing and is driven by rods hung from a crosshead on the end of the piston rod. The merit here was a considerable saving in headroom and the type found favour with marine engineers. The PS *Comet*, the first steamboat to run commercially in Europe, had an engine built by John Robertson of Glasgow in 1812, and magnificent examples were turned out later by Denny of Dumbarton and John Penn of Greenwich for both passenger vessels and warships. The type was not superseded until the middle of the century.

The most important development of all to come with the new century was high-pressure steam, bringing major increases in power and improvements in economy. The engineer responsible was Richard Trevithick (1771-1833), who devoted many years of research to the principle and invented, among many other things, what came to be known as the 'Cornish boiler'. The old 'haystack' and 'wagon' boilers of copper and lead or wrought iron, with numerous leaks which the boilerman would try to patch with tallow and tow, vanished almost at once; the new design was cylindrical, of stout wrought-iron plates riveted together with a fire grate-cum-flue constructed down the middle. In 1801 Trevithick had constructed a successful road engine, to be followed in 1804 by the first steam railway locomotive. The new Cornish boiler was ideally suited for mounting on wheels.

Steam at 50 pounds per square inch was

TOP: *A distinctive little design, the beam supported on a single column with four struts, was installed at Bagge's Brewery, Norwich, in 1840. This engine has been in the Norwich Museum since 1930.*

LEFT: *About 1838 and for some years after, J. & E. Hall of Dartford produced a remarkably successful little compound beam engine, self-contained and mounted on a tank-frame. This one was at Sittingbourne but is now at the Henry Ford Museum, Dearborn, USA.*

now available to designers, and one of the first to take advantage of it was another Cornishman, Arthur Woolf (1776-1837), who invented compounding — that is, steam being used first of all in a high-pressure cylinder and then expending its residual force in a larger, low-pressure cylinder working the same beam.

Economies of the order of fifty per cent were found to be obtainable if the design was right, but the proportional sizes of the two cylinders proved to be critical. Engine-building had by now moved right out of the sphere of the blacksmith and carpenter to involve new trades such as those of the mechanical engineer and steam fitter.

THE CORNISH ENGINE

The rotative beam engine, with its connecting rod, crank and flywheel, was now well established and in increasing demand for driving the mills and factories which were springing up all over the country. This, however, by no means spelled the end of the older reciprocating type of beam engine which came into greater and greater use for pumping out mines and providing domestic water supplies. One of the main industrial areas of England was Cornwall, where there was an unprecedented demand

for tin and copper as well as tungsten and various other metals. Lead was being worked everywhere on the Mendips and Pennines, in southern Scotland and in every county of Wales. New coal-mines were continually being sunk. All these holes in the ground needed to be kept free of the water which poured into them from the subsoil. Firms were established in many industrial areas to manufacture beam pumping engines, but most of all in Cornwall, where enormous foundries turned out at

BELOW: *A pair of 30-inch rotative engines at Basset Mines in Cornwall driving a battery of sixty tin-stamps, used to reduce the ore to powder before refining.*

RIGHT: *The Science Museum model of Smeaton's improved atmospheric engine installed at Long Benton Colliery, Newcastle, in 1772.*

short notice steam engines, boilers, pumping equipment and pitwork of every possible description. The largest of them all, Harvey's of Hayle, included shipbuilding in its activities and was capable of constructing engines with cylinders of up to 144 inches in diameter. Other factories of note which supplied many hundreds of engines in fierce competition were the Perran Foundry and the Copperhouse Foundry, but there were many others only slightly smaller.

The so-called Cornish cycle was similar to the single-acting principle established by Watt, with steam acting on top of the piston but using the steam expansively, thus producing enormous fuel savings.

The typical Cornish engine had its huge beam mounted on the strengthened end wall of the engine-house so that the mechanism was indoors and the end of the beam, with pumprods extending down the mineshaft, was in the open air. The pumprods were composed usually of Oregon pine, in section about 18 inches square and often weighing, in the case of a deep mine, up to a hundred tons. The first pump was in a sump at the bottom of the mine and the water was raised in lifts with intermediate tanks and pumps, worked from offsets on the same pumprods, every few hundred feet. It was not necessarily pumped to the surface, but often to a tunnel or adit driven to come out on lower ground, often miles away. West Cornwall is still riddled with such tunnels, and the construction of them with hand tools was a staggering achievement.

What the engine did was to raise the pumprods and, with them, the plungers in the pumps, thus filling them with water. When the steam valve shut, the rods descended again under their own great weight, an 'equilibrium valve' opening at the right moment to allow the expanded steam to fill the space below the piston as it rose. Closure of this valve as the piston approached the top of its stroke cushioned the descent of the rods, bringing them gently to rest so that the pumps could take the water at their own natural speed. This valve was invariably known to the enginemen of old as the 'Uncle Abram' valve. The engine then paused at the top of its stroke until the cataract allowed the steam valve to open and repeat the cycle. The valves were adjustable to cope with whatever quantity of water needed pumping out, and it used to be said by the engineers that a good engine could work at ten strokes a minute or one stroke in ten minutes. To the miners in the gloomy, candlelit galleries far below ground, the sound of the pump echoed for miles, sounding like a giant snore and giving them constant assurance that all was well.

Engines working on the Cornish cycle could also be made rotative by using a heavy connecting rod, known as a 'sweep rod' to induce half the crank's rotation by gravity. But of the many rotative engines used in Cornish mines, the majority were double acting with the valves operated from a plug rod or the crankshaft.

Rotative beam engines used to hoist the ore were called 'fire whims' to distinguish them from the old horse whims used previously. They were also used to drive batteries of stamps for crushing the ore or for pumping lesser quantities of water. Some stamp engines and whims also drove pumps and had a second beam projecting from the rear of the house for this purpose. Another quaint device which took its drive from the rotative engine was a 'man engine' employed in some deep mines to move men up or down the shaft. In this a vertical timber rod with small platforms and handles attached at fixed intervals was driven up and down by a crank at steady speed, the platforms coming level with fixed platforms or 'sollars' in the shaft at the top and bottom of each stroke. By stepping on and off during the momentary dwell between strokes, men could travel up or down.

Design in Cornwall was stimulated by a monthly publication detailing the duty of all contributing engines, an arrangement which produced intense rivalry between designers and enginemen during the heyday of metal-mining. This was in the 1840s and 1850s, and it is believed that by mid century there were about two thousand steam engines at work in Cornwall, besides many more from Cornish makers in other parts of Britain and most mining districts in other countries. The 1860s, however, brought a bad slump and the demand for equipment plummeted. Piped water supplies were being provided by the larger cities at that time and the Cornish

ABOVE: *At Kew Bridge pumping station there is a 90-inch Cornish engine installed in 1846, which worked for 97 years. It was built by Sandys, Carne and Vivian at the Copperhouse Foundry, Hayle, and each of its 11-foot strokes delivered 472 gallons. Now it runs for the public at weekends.*

RIGHT: *The only remaining example of a marine side-lever engine used in a water-pumping station is at Turnford, Herts. The maker was Boulton, Watt & Co. (1848). It has a 28-inch piston.*

RIGHT: *Coalbrookdale, Salop, was the cradle of the Industrial Revolution. Here, at the Blist's Hill Museum, is the great double engine, David and Sampson, which pumped the air for the blast furnaces and was known as a 'blowing engine'. It was built in 1851 by the Lilleshall Company and functioned for a hundred years.*

BELOW: *The engine at Combe, Oxfordshire, powered a sawmill for sixty years, after which it was idle for another sixty before being brought back to life by a team of enthusiasts. Engine and boiler, by Thomas Piggott of Birmingham, date from 1852. The 18-inch diameter cylinder is double acting.*

foundries came into competition with firms from the Midlands and North for providing pumping equipment. However, new Cornish pumping engines continued to be built and second-hand engines overhauled and re-erected until the 1920s, and examples continued at work into the 1960s.

A self-contained compound engine typical of practice in the 1860s. This one was made by Thomas Horn of Westminster for powering small mills and workshops and has the single fluted column typical of its day. This example was presented to the Science Museum by the Newcomen Society.

THE HEYDAY OF STEAM

Cornish engines were, of course, made by many firms outside the Duchy, just as Cornwall produced all manner of other engine designs. It was not until the mid nineteenth century that direct-acting horizontal engines of various designs, probably as a result of railway-locomotive development, began to challenge the beam-engine layout. By 1840 it did not seem that much greater efficiency could be wrung from steam engines (although considerable improvements in boilers had still to come), but designs continued to be delightfully varied; designers were still able to be rug-

ged individualists and detail improvements took all sorts of unrelated forms.

Until about that time, engines had been severely plain in construction, but by mid century there were tentative efforts to dignify or even to beautify them, and with the onset of the 1860s the outward appearance of the large rotative designs had radically changed. Cornwall had no part in this fashion, but other makers produced to order 'Egyptian' engines, with lotus-headed columns and governors shaped like winged scarabs, or Greek motifs, the most popular of all, with

At Wollaton Hall, Nottingham City Museums have re-erected one of a pair of compound rotative beam engines by Hawthorn of Newcastle (1858). Originally sited at Basford water pumping station. The engine developed 34hp at twelve strokes a minute and 40 pounds per square inch steam pressure. Note the Cornish-type valve lifting gear.

graceful, fluted Doric columns, often painted in dark red, and a profusion of openwork. These big and beautiful machines were a source of commercial or civic pride, and chairmen took care that their engineering installations were admirable as well as efficient — which is as it should be. Engine-houses, too, were carefully and effectively designed with a monumental or even cathedral-like air, and many of these remain, even where the faithful engines have been ruthlessly scrapped after a hundred years of service.

A few beam engines are still at work, doing the job they were designed for. Others are being restored and steamed by groups of devoted amateurs. Yet others are perhaps still being scrapped, but with the remarkable increase of interest in Britain's engineering heritage and the growth of tourism scrapping seems to be coming to an end and there is a vigorous movement to restore more engines to steam. This is not before time, since many unique designs have been lost for ever. The larger examples are some of the most impressive of all man-made objects, and nobody who has ever seen one of them at work and has climbed two sets of stairs to watch a 40-ton beam rising and falling 11 feet in the air in almost perfect silence is ever likely to forget the experience.

ABOVE: *All that remains of the Hammersmith pumping engine, made by Harvey's in 1854, is a cast-iron beam fabricated as one casting. It is the only remaining openwork beam in Britain and has been strengthened with a bridle.*

BELOW: *This engine from Llanishen Reservoir, South Glamorgan, is a single-column pumper built by Harvey's of Hayle in 1851. It supplied a million gallons a day at 30 rpm and is now at the Welsh Industrial Museum, Cardiff.*

BELOW: *The Matthews Weaving Shed engine, built by Bracewell of Burnley in 1861, has an extra cylinder (added on the McNaught system) behind the connecting rod on the left. The engine has been re-erected by the Bradford Museums.*

ABOVE: *The pumping engine at Prestongrange Colliery, East Lothian, was made by Harvey's in 1874 and has a 70-inch cylinder with 12-foot stroke.*

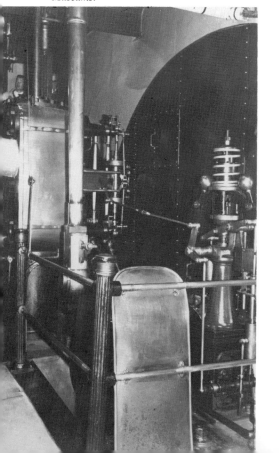

ABOVE: *The piston rods and parallel motion of one of the four great Gimson compounds installed in 1891 to pump the sewage of Leicester. The station is now a museum of technology.*

ABOVE: *At Lound, Suffolk, pumping station are preserved a pair of 1857 'grasshopper' engines by Easton, Amos and Anderson. Instead of a central pivot, the beams are supported at one end on a swinging link.*

BELOW: *Coleham sewage pumping station is now one of Shrewsbury's museums. It houses a pair of Renshaw compound rotative engines installed in 1900 and in continuous operation until 1970.*

ABOVE: *Papplewick pumping station supplied Nottingham with water from 1884 until 1969. The James Watt & Co. engines are single-cylinder of 170 hp each, with 20-foot flywheels and beams 27 feet long.*

PLACES TO VISIT

MUSEUMS WHERE BEAM ENGINES CAN BE SEEN

Intending visitors should contact museums for information on steaming days.

Birmingham Museum of Science and Industry, Newhall Street, Birmingham B3 1RZ. Telephone: 021-236 1022.

Bolton Mill Engine Museum (Northern Mill Engines Society), Atlas Number 3 Mill, Chorley Old Road, Bolton, Lancashire.

Bradford Industrial Museum, Moorside Road, Eccleshill, Bradford BD2 3HP. Telephone: Bradford (0274) 631756.

Bressingham Gardens and Live Steam Museum, Bressingham, Diss, Norfolk IP22 2AB. Telephone: Bressingham (037 988) 386.

Bridewell Museum, Bridewell Alley, Norwich, Norfolk NR2 1AQ. Telephone: Norwich (0603) 611277, extension 299.

British Engineerium, Nevill Road, Hove, East Sussex BN3 7QA. Telephone: Brighton (0273) 559583 or 554070.

Calderdale Industrial Museum, Central Works, Halifax, West Yorkshire. Telephone: Halifax (0422) 59031.

Coleham Pumping Station, Longden Coleham, Shrewsbury, Shropshire. Telephone: Shrewsbury (0743) 61196.

Eastney Pumping Station, Henderson Road, Eastney, Portsmouth PO4 9JF. Telephone: Portsmouth (0705) 827261.

Forncett Industrial Steam Museum, Hannays, Forncett St Mary, Norfolk.

Greater Manchester Museum of Science and Industry, Liverpool Road, Manchester M3 4JP. Telephone: 061-832 2244.

Hunday National Tractor and Farm Museum, Newton, Stocksfield, Northumberland. Telephone: Stocksfield (0661) 842553.

Ironbridge Gorge Museum, Ironbridge, Telford, Shropshire TF8 7AW. Telephone: Ironbridge (095 245) 3522.

Kew Bridge Engines Trust and Water Supply Museum, Kew Bridge Road, Brentford, Middlesex. TW8 0EF. Telephone: 01-568 4757.

Leeds Industrial Museum, Armley Mill, Canal Road, Armley, Leeds. Telephone: Leeds (0532) 637862.

Leicestershire Museum of Technology, Abbey Pumping Station, Corporation Road, Abbey Lane, Leicester. Telephone: Leicester (0533) 661330.

Newcomen Engine House and Dartmouth Museum, The Butterwalk, Dartmouth, Devon TQ6 9PZ. Telephone: Dartmouth (080 43) 2923.

Nottingham Industrial Museum, Courtyard Buildings, Wollaton Park, Nottingham NG8 2AE. Telephone: Nottingham (0602) 284602.

Poldark Mine, Wendron, Helston, Cornwall. Telephone: Helston (032 65) 4549.

Royal Scottish Museum, Chambers Street, Edinburgh EH1 1JF. Telephone: 031-225 7534.

Ryhope Engines Museum (Trust), Ryhope, Sunderland, Tyne and Wear. Telephone: Sunderland (0783) 210235.

St Austell China Clay Museum, Wheal Martyn Museum, Carthew, St Austell, Cornwall PL26 8XG. Telephone: St Austell (0726) 850362.

Science Museum, Exhibition Road, South Kensington, London SW7 2DD. Telephone: 01-589 3456.

Scottish Mining Museum, Pretongrange, Prestongrange, East Lothian EH32 9PX. Correspondence to: the Director, Scottish Mining Museum, Lady Victoria Colliery, Newtongrange, Midlothian EH22 4QN. Telephone: 031-663 7519.

Somerset County Museum, Taunton Castle, Castle Green, Taunton, Somerset TA1 4AA. Telephone: Taunton (0823) 55504.

Strumpshaw Hall Museum, Strumpshaw, near Norwich. Telephone: Norwich (0603) 714535.

Welsh Industrial and Maritime Museum, Bute Street, Cardiff. Telephone: Cardiff (0222) 481919.

ACCESSIBLE ENGINES

The following engines can usually be seen at reasonable times: most of them are in public places.

Bedford: waterworks engine. Mander College Grounds, Cauldwell Street.

Birmingham: Grazebrook blowing engine on roundabout at south end of A38M Aston Expressway.

Dumbarton: Napier side-lever engine in the public park.

Loughborough University, Leicestershire: a Watt engine of 1850 at the main entrance.

Soudley, Gloucestershire: Hewlett's engine on Camp Mill Site.

Stretham, Cambridgeshire: land drainage engine just east of A10. 5 miles south of Ely (apply at any reasonable hour).

Wanlockhead, Dumfriesshire: hydraulic engine in the village.

ENGINES OPERATED BY TRUSTS OR CLUBS

Full details of opening and steaming times should be sought before making a visit.

Broken Scar Waterworks, Tees Cottage Pumping Station (Trust), Coniscliffe Road, Darlington, County Durham. Telephone: Darlington (0325) 2869.

The Canal Pumping Station, Claverton, near Bath, Avon. Telephone: Bristol (0272) 712939.

Combe Sawmill (Combe Mill Society), Combe, near Bladon, Oxford. Telephone: Woodstock (0993) 811118.

Crofton Pumping Station (The Crofton Society), near Great Bedwyn, Wiltshire: two Cornish engines. (A branch of the Kennet and Avon Canal Trust.) Telephone: Devizes (0380) 71279.

Dogdyke Pumping Station (Trust), Bridge Farm, Tattershall, Lincolnshire.

Leawood Pumphouse, Cromford Canal Scoiety, Cromford, Derbyshire. Telephone: Wirksworth (062 982) 3727.

The Levant Engine, St Just, Cornwall. The engines at East Pool, Camborne, are in working order and open except in winter. For details apply to the National Trust Cornwall Office, Lanhydrock, Bodmin, Cornwall PL30 4DE. Telephone: Bodmin (0208) 4281 or 4284.

Middleton Top Engine House, Middleton, Derbyshire. Telephone: Wirksworth (062 982) 3204.

Papplewick Pumping Station (Papplewick Trust), Longdale Road, Ravenshead, Nottingham. Telephone: Nottingham (0602) 46651.

Pinchbeck Marsh Pumping Station (Welland and Deepings Internal Drainage Board), Welland Terrace, Spalding, Lincolnshire PE11 2TD. Telephone: Spalding (0775) 5861.

Westonzoyland Pumping Station (Trust), Westonzoyland, near Bridgwater, Somerset. Telephone: Taunton (0823) 412713.

ENGINES WHICH CAN BE VISITED BY SPECIAL ARRANGEMENT

In some cases parties only are acceptable. Intending visitors should write first.

Blagdon Pumping Station, Avon. Bristol Waterworks Company, PO Box 218, Bridgwater Road, Bristol.

Clay Mills Pumping Station, Burton upon Trent, Divisional Manager, Severn Trent Water Authority, Trinity Square, Horninglow Street, Burton upon Trent, Staffordshire DE14 1BL.
Dalton Pumping Station, Cold Hesleton, Seaham. Sunderland and South Shields Water Company, 29 John Street, Sunderland SR1 1JU.
Henwood Pumping Station, Ashford, Kent. Divisional Manager, Mid-Kent Water Company, Wallis Road, Ashford, Kent.
Hopwas Pumping Station, Tamworth, Staffordshire. South Staffordshire Waterworks Company, Central Office, Green Lane, Walsall WS2 7PD.
Lound Pumping Station, Lowestoft. East Anglian Water Company, 163 High Street, Lowestoft, Suffolk.
Markfield Road Sewage Pumping Station, Tottenham, London. River Lea Archaeological Society, c/o Lea Valley Regional Park Authority, Myddleton House, Bulls Cross, Enfield, Middlesex EN2 9HG.
Mining Department, Heriot-Watt University, Edinburgh EH1. Telephone: 031-225 8432.
Roall Waterworks, Whitley Bridge, Goole. The Director, South East Division, Yorkshire Water Authority, Copley House, Waterdale, Doncaster, South Yorkshire.
Sandfields Pumping Station, Chesterfield Road, Lichfield, Staffordshire. Apply as for Hopwas Pumping Station.
Springhead Pumping Station Museum, Yorkshire Water Authority, Eastern Division, Hull. Telephone: Hull (0482) 28591.
Turnford Pumping Station, Canada Lane, Great Cambridge Road, Turnford, Wormley, Hertfordshire. Divisional Manager, Thames Water Authority, 173 Rosebery Avenue, London EC1R 4TP. Telephone: 01-681 7131.
The Whitelees Beam Engine, Holcroft Castings and Forgings Limited, Whitehall Street, Rochdale. Telephone: Rochdale (0706) 40911.

Further details of the above and other engines are available in a leaflet *Industrial Archaeology and the Water Industry* (WASC Leaflet Number 6). Send sae to the Water Space Amenity Commission, 1 Queen Anne's Gate, London SW1H 9BT.

FURTHER READING

Barton, D. B. *The Cornish Beam Engine.* D. Bradford Barton, 1966.
Bird, R. H. *Yesterday's Golcondas.* Moorland, 1977.
Crowley, T. E. *The Beam Engine.* Senecio Press, 1982.
Dickinson, H. W. and Jenkins, Rhys. *James Watt and the Steam Engine.* 1927. (Reprinted Moorland, 1981).
Hayes, Geoffrey. *A Guide to Stationary Steam Engines.* Moorland, 1981.
Hayes, Geoffrey. *Stationary Steam Engines.* Shire Publications, 1979.
Hodge, James. *Richard Trevithick.* Shire Publications, 1978.
Law, R. J. *The Steam Engine.* HMSO (Science Museum booklet), 1965.
Rolt, L. T. C. *James Watt.* Batsford, 1962.
Rolt, L. T. C. *Thomas Newcomen.* David and Charles. 1963.
Rolt, L. T. C. and Allen, J. S. *The Steam Engine of Thomas Newcomen.* Moorland, 1977.
Vale, Edmund. *The Harveys of Hayle.* D. Bradford Barton, 1968.
Watkins, George. *The Stationary Steam Engine.* David and Charles, 1968.
Woodall, F. D. *Steam Engines and Water Wheels.* Moorland, 1975.

ACKNOWLEDGEMENTS
Acknowledgement for the use of photographs is due to the following. Crown copyright, Science Museum, London, pages 4, 9, 13, 15 (both), 21, 25; Photo, Science Museum, London, pages 6-7, 8, 17 (bottom), 19 (bottom); North Western Museum of Science and Technology, page 11; John Day, Dorking, pages 16 (right), 19 (top), 24 (top), 28 (bottom left); National Museum of Wales, pages 17 (top), 27 (bottom); Royal Institution of Cornwall, page 20; Peter Floyd, *Oxford Mail and Times,* page 24 (bottom); C. E. Lloyd, Beeston, page 26; Museum of Technology, Leicester, page 28 (bottom right). Prints were also most generously provided as follows: S. A. Staddon, page 1; Thames Water Authority, pages 2, 3, 4, 23 (both); National Coal Board, South Yorkshire Area, page 16 (left); Derby Evening Telegraph, page 18; Prestongrange Open-air Museum and David Spence, page 28 (top); R. L. Greensmith, page 30 (top); Ian Taylor, page 29 (bottom). The publishers acknowledge with gratitude the assistance of Mr Kenneth Brown and are especially indebted to Mr. J. H. Andrew, Keeper (Science), Birmingham City Museums and Art Gallery, for advice on the text and for his photograph of the Amos engine on the front cover, reproduced by courtesy of Birmingham City Museums and Art Gallery.